REVIEWS FOR THE AI-POWERED ONE HOUR WORKDAY

★★★★★ **K.T.** 5.0 out of 5 stars **Not an AI Expert? Good. This Book Is Still for You.**

Reviewed in the United States on January 26, 2026 - Verified Purchase

I listened to The AI One Hour Workday twice and I genuinely loved it. First of all, Nicole McCain is an excellent new writer. The way she uses real people's experiences and case studies makes the book feel less like a "business book" and more like an intervention… because that's exactly what it was for me.

This book will show you who you really are in the mirror, even if you don't love what you see at first.

And I'm saying that as a woman who is doing the absolute MOST. Entrepreneur, content creator, trying to build and grow a new website, managing another brand with a VA, handling life as a solo household… and I foster dogs and help neglected/surrendered animals get to safety. When a rescue is out of space, I don't get to say "not today." That becomes a judgment task, meaning it's something only I can do in that moment.

One section in the book really grabbed me: Nicole has you list your tasks as Judgment vs. Memory. Judgment = only you can do it. Memory = ongoing, consistent tasks that can be systemized or delegated. My list was… shockingly… mostly Judgment. It didn't even sound real until it was sitting there on paper staring back at me.

AF854344

And here's the part that surprised me: I couldn't even complete the 48-hour audit the way I wanted to.

Not because the method doesn't work, but because my life right now is so full that I kept getting disrupted every single day by something urgent that required me. Attorneys. Interviews. Home maintenance. Dogs recovering from surgery. Cleaning. Cooking. Eating (or forgetting to). And yes… my coffee staying cold before I ever get the first sip.

That's when I realized… I didn't need "more motivation."

I needed structure. I needed systems. I needed an actual plan to stop living inside chaos.

This is why I want to make something clear: you do NOT need to understand AI to get value from this book. Even if you're completely clueless to AI, this book is still for you. It's for anyone overwhelmed, overextended, and trying to do more than one human being is meant to do in a day.

The workbook alone is worth the investment. Nicole gives you both an interactive digital workbook and a printable version, and it helps you see exactly where your time is going and why you feel drained. The 48-hour audit is simple and powerful: track what you do for 48 hours, how long it took, your focus level (low/medium/high), interruptions, and whether it repeats. That exercise will humble you in the best way.

If you're in my shoes and the audit feels impossible to start, my recommendation is this:

Start messy. Start with ONE day. Start with notes in your phone. Voice record it if you need to. Don't wait for a "calm week," because calm weeks don't magically appear when you're building a life and a business at the same time. The point is progress and awareness, not perfection.

This book helped me step back, declutter my mind, turn off the apps that demand immediate attention, and stop volunteering my time in ways that leave me burned out. I'm now focused on building my Execution Map, prioritizing from a clean slate, and committing to my 30-day focus, because I refuse to keep running a business and a life that only works when I'm running on straight up adrenaline.

Bottom line: If you're a female CEO, entrepreneur, content creator, or anyone trying to do it all, this book is a must. It's not just about AI… it's about getting your life back.

Highly recommend. Thank you, Nicole McCain, for the intervention I didn't know I needed.

★★★★★ **Robert F.** 5.0 out of 5 stars **Want to REALLY use AI to make your business run better? START HERE.** Reviewed in the United States on January 28, 2026. - Verified Purchase

If you're done looking for AI hacks to make running your business easier, The AI-Powered One Hour Workday is THE BEST place for you to start. Nicole McCain presents an overarching strategy to help you integrate AI tools and processes into your business that will streamline operations, save time, and reduce stress and let you focus on what really moves the needle. It's all about preserving that "One Thing" that only you can do and which must be done, rather than racing from one fire to the next that shouldn't be happening in the first place. The "HAS" framework is exquisite in it's simplicity for providing structure, yet flexible enough to apply across any aspect of your business, and she illustrates its use with clearly described examples throughout the text.

Whether she is explaining how to trace your workflows throughout the business with the 48-Hour Truth Audit(tm), setting up your AI Brain, working through the decision-logic of the HAS framework, of simplifying the follow-up to measure success, Nicole provides ample explanation not only of the techniques but also the important "WHY" of each step in the process. Additionally, if you enjoy a good mindset shift, Nicole provides just the right amount to get you fired up without feeling like you're stuck at the Kool-Aid stand at Cheerleading Camp.

The AI-Powered One Hour Workday is a quick read that will get you oriented toward more strategic, and powerful, use of AI for your business. If you add the cost of the book, and the hour or two to read it, you'll get FAR MORE than 100% ROI in the end. Try Nicole's approach once and you'll find success. Incorporate her approach into the way you build your business and you'll learn what true "winning" looks like.

★★★★★ **Taneya M. Taylor** 5.0 out of 5 stars **Strong AI Productivity Framework—Best With Clear Systems** Reviewed in the United States on January 15, 2026 - Verified Purchase

AI One-Hour Workday offers a refreshing perspective on how AI can dramatically reduce busywork and help you focus on high-leverage tasks. The book does a solid job of reframing work around output, systems, and decision compression rather than long hours and constant hustle.

What I appreciated most is that it treats AI as a force multiplier, not magic. When applied to the right workflows, the strategies can genuinely save time and mental energy.

That said, the “one hour” framing feels more aspirational than literal. The results depend heavily on having clarity, discipline, and well-defined processes before applying AI. Without that foundation, readers may struggle to see the promised benefits.

Overall, this is a valuable read for entrepreneurs, consultants, and knowledge workers who want to work smarter but it works best when paired with strong decision-making and intentional systems.

★★★★★ **Tonya** 5.0 out of 5 stars **Makes You Want to Take Action Immediately**

Reviewed in the United States on January 19, 2026

The AI Powered One-Hour Workweek is an easy, motivating read that makes AI feel simple instead of overwhelming. It’s great for beginners and still valuable for more experienced readers because it shows how small, practical changes can be implemented the same day and immediately save time.

THE AI-POWERED ONE HOUR WORKDAY

How Business Owners Use AI to Reclaim Time, Reduce Chaos, and Build a Business That Runs Without Them

Nicole McCain

The AI One Hour Workday

The AI One Hour Workday™

The AI-Powered One Hour Workday™

ISBN: 979-8-9945094-1-8

For permissions, contact:

Nicole McCain nicole@theaionehourworkday.com

Printed in the United States

First Printing 2026 First Edition 2026

DISCLAIMERS:

This book provides general information about using artificial intelligence tools in business operations. It is not legal, financial, tax, or professional advice specific to your situation. The author and publisher are not liable for any business or personal decisions, financial losses, or other outcomes resulting from applying the information in this book.

AI tools, software, and platforms change frequently. Features, pricing, and availability mentioned in this book were accurate as of the publication date but may have changed. Readers should verify current information before implementing any recommendations.

The author makes no guarantees about the results you'll achieve by reading this book. All business requires risk and hard work. The results mentioned are examples and not guarantees. Individual results will vary based on business type, implementation effort, and numerous other factors.

Trademarks and Third-Party References:

All trademarks, personal names, product names, and company names mentioned in this book are the property of their respective owners. The use of these names do not imply any endorsement, affiliation, or sponsorship.

ChatGPT, Google, Claude, Notion, n8n, GHL, and other related names referenced in this book belong to their respective owners.

DEDICATION

To my wonderful family, especially my two little girls, for believing in me and cheering for me all the way. Telling me, "Mommy, you can do it!" The sparkle in your eyes and excitement when you see mommy keeps me going and makes anything in this world possible.

May God continue to be a lamp onto your feet and light onto your path.

TABLE OF CONTENTS

INTRODUCTION

There was a time in my business when I forgot what ease felt like.

I remember sitting at my desk late one night. The house was quiet. The laptop screen felt too bright. I stared at a task list that somehow kept growing, even though I kept crossing things off. Client messages waited for replies. Content deadlines stacked up. Notes need organizing. Ideas had nowhere to land. Systems lived across tools and my own brain.

On paper, everything looked fine. In my body, it felt heavy.

Every day felt full, but not fulfilling. Every night ended the same way. I did a lot today, but I couldn't clearly name what actually moved.

If you've ever felt like your business only runs because you're constantly holding it together, you know this feeling.

It took me longer than it should have to see the pattern. The work itself wasn't the problem. The problem was how much the work depended on me. My business required me to remember, respond, follow up, and keep track of everything. That weight didn't arrive all at once. It built quietly, day after day, until being stretched thin felt normal.

A Direct Conversation Before We Go Further

When you see the phrase "one-hour workday", your reaction is probably immediate.

It might sound unrealistic. It might feel like it applies to someone else. It might trigger skepticism for very real reasons.

That reaction usually comes from your own experience.

You're not running a hobby and you're not new to responsibility. You've never been confused about effort and you're running something real. There are clients, expectations, and consequences when things slip. This book was written with that reality in mind.

What the One Hour Actually Refers To

Throughout this book, the one-hour workday has a very specific meaning.

It refers to the time your business truly needs you to step in and do the work that only you can do. It is the time required for work that depends on your unique judgment, your decisions, and your ability to execute. This includes providing your voice, your perspective, solving real problems, and doing work that changes outcomes. It is the work that cannot be copied, delegated, or rushed without consequences.

The AI One Hour Workday is designed to protect this time. It reduces the number of decisions you have to make, shortens how long execution takes, and removes the follow-up work that usually drags on after a task is finished. Systems handle the repeatable parts of your business whether you are available or not. That return of time creates more free space in your mind and calm in your day.

Ask yourself this question each morning:

What is the one critical thing I must personally do today that will move my business forward?

According to the AI One Hour Workday principles, it should not include low-value tasks that simply need to happen. Things like follow-ups, admin, repeated explanations, searching for context, restarting tasks after interruptions, or holding loose ends in your head.

When those things are handled by structure instead of manual effort and memory, your time changes shape. Your business still runs, your clients still feel supported, and progress does not stall. You are no longer acting as the glue holding everything together but instead, you are directing the work that actually matters.

Why the Most Valuable Work Requires Space

I've always paid attention to how some founders work fewer hours while earning far more than the average business owner. The difference has nothing to do with discipline or stamina. It comes down to what kind of work they protect space for.

Warren Buffett is a clean example. He spends much of his day reading, thinking, and making a small number of decisions that actually move outcomes. His calendar stays intentionally light because his value isn't in staying busy; it's in judgment. Some have claimed that Buffet only works 2 hours a day!

Bill Gates once said that seeing Buffett's mostly empty calendar changed how he thought about time. Gates was used to filling every minute with work while Buffett protected his space. That space is where the real "work" happens.

Every time I come back to that, I'm reminded that the real question isn't how much more we can do. It's what low-return work we're willing to let go of to create that kind of mental space for ourselves. The space that will allow us to use our time and brain to come up with new, creative ideas, concepts, or even those billion dollar strategies that Buffet was able to come up with.

Why the Workday Gets So Heavy

Most business owners start with good intentions. They want to build something meaningful. They want flexibility. They want work that fits into a life.

Over time, flexibility turns into constant availability. Meaningful work gets crowded out by reminders, manual effort, and constant context switching. Each day stretches longer without actually creating space.

Research from Slack and Salesforce shows that small business owners lose close to ninety-six minutes a day to wasted time. That loss comes from distractions, waiting on updates across tools and people, and constantly switching between apps.

I've experienced this first hand in the various jobs I held. Everyone has to wait for the decision maker or a sign off before things can move forward. These types of dependencies lead to lost hours, low morale and slow progress.

When a business relies on one person to move everything forward, the pressure becomes constant. That pressure shows up as mental load, background stress, and the feeling that you can never fully shut off.

I wrote this book because I know what it feels like to want more presence in your life while your business keeps pulling you back. I know what it's like to hear your kids in the other room calling for you while you promise yourself you'll be done soon. Those moments carry more guilt than we like to admit.

What This Book Helps You Build

This book offers a different way to relate to your work. It helps you see your work patterns clearly for the first time. Not the version you think your days look like, but the reality of where your time, attention, and energy are actually going. Once those patterns are visible, you can finally decide what deserves your focus and what never should have been on your plate.

You'll learn how to prioritize handing off manual, repeat work so your days stop getting eaten by low-value tasks. Instead of being pulled in every direction, you'll learn how to protect focused, uninterrupted time for the work that actually requires you.

AI shows up here as support, not control. It carries repeat work, holds context, tracks details, and keeps things moving so you're not constantly restarting, re-remembering, or cleaning up loose ends. Your judgment, creativity, and decision-making stay with you.

Inside these pages, you'll learn how to identify which tasks can be handed off, how to design workflows that reduce mental load, and how to apply deep, focused work to the moments that matter most. You'll learn how to use AI to support thinking, delivery, and follow-up without diluting your voice, and how to measure progress so the relief you feel is real and visible.

The goal is a business that supports you instead of leaning on you. When that shift happens, your days feel different. Your focus sharpens. Your mind gets quieter. Your energy goes toward work that actually moves things forward.

Who This Book Was Written For

This book is for founders, consultants, coaches, creatives, and business owners who want their business to fit inside their life. It's for people who care deeply about quality and want more control over their time.

You'll recognize yourself here if you work long hours and still feel behind, if you feel mentally on call even when you stop working, if you know there's a better way but don't know where to start, or if you want to use AI thoughtfully.

This book will frustrate you if you're looking for a catalog of AI tools, step-by-step tutorials, or instant results without effort. It also won't land if you're unwilling to rethink how you currently work or if you believe doing everything yourself is the best option.

How the Book Is Structured

This book is designed to meet you where you are and walk you forward in a clear sequence.

We begin by slowing things down enough to see the truth about how your workday really operates. Not the ideal version. Not the planned version. The real one. You'll identify patterns, hidden work, and the quiet drains on your time and attention that have been shaping your days.

From there, we map how work actually moves through your business. You'll see where things depend on your memory, where effort is being duplicated, and where manual steps are creating unnecessary pressure. Once that structure is visible, the decisions become obvious.

Only then do we introduce AI and automation. Not as tools stacked on top of chaos, but as support systems designed to carry repeat work, hold context, and keep things moving without constant supervision. You'll learn how to protect deep, focused work for yourself while systems handle what never needed your judgment in the first place.

Each section builds on the last. Nothing is rushed. Nothing requires technical expertise, a large team, or a full rebuild. You're not adding more to your plate. You're removing what never should have been there.

The goal is practical and personal. More mental space. More consistency. More control over how your time is spent. The ability to serve your clients well without being constantly on call, and the freedom to step away without worrying that everything will stall the moment you do.

Before We Begin

If you feel cautious right now, that's okay.

This book doesn't ask you to believe in anything. It asks you to become more aware.

My hope is that these pages offer something rare. A path back to yourself. A business that supports your life instead of consuming it. An identity where your time is protected, your energy is respected, and your brilliance isn't buried under tasks that never should have been yours.

If you take these steps, even small ones, your world will shift. Not because of the tools, but because of how you begin to see your business and yourself.

Thank you for choosing ease over chaos. Thank you for believing your time is worth protecting.

You're not only building a lighter business but instead, you are building a life with more room for you in it.

So here's my only question before we begin:

Are you willing to believe that your workday doesn't have to feel like a sprint you never signed up for?

If the answer is even a small yes, a quiet yes, a tired yes, a skeptical yes, then let's begin.

Your lighter business starts here.

PART I

WHY YOUR BUSINESS FEELS HEAVY

(EVEN WHEN YOU'RE DOING EVERYTHING RIGHT)

CHAPTER 1

WHY EFFORT STOPPED WORKING BEFORE YOU NOTICED

Close your eyes for a moment.

Imagine waking up on a weekday morning, and nothing feels urgent. Your inbox is under control. Your email and content are already drafted and scheduled. Yesterday's meeting notes have been turned into clear action steps. You ease into the day because your business is already moving without you pushing it.

Now open your eyes.

If I took your laptop right now, would this feel like your reality? And at the end of most days, could you clearly name the three things you personally did that directly impacted revenue?

Not tasks you completed. Not fires you put out. Actual progress. Things like streamlining a delivery process, moving an offer forward with a potential client, or making a decision that unlocked growth.

And after everything else you handled today, do you still have the mental bandwidth to create new offers, out of the box solutions, or build strategic partnerships?

Most business owners usually don't until they go on vacation. Their days get sliced into tiny pieces.

Have you ever finally had the time to sit down to think, but by then your best energy was already gone?

That was the feeling I couldn't ignore on a Tuesday morning when everything finally clicked for me.

I was at my desk with a cup of coffee that had gone cold, staring at a to-do list that seemed to grow longer by the second. The kind of list that looks reasonable when you run a business.

Draft this. Fix that. Check this. Respond to that. Update this. Review that.

None of it was hard. None of it required my best thinking. And yet it felt heavy. I kept scrolling and thinking, why does this feel like I'm working just to stay in place?

Then my phone buzzed. A client message came through. "Have you reviewed my strategy documents yet?"

I already knew the answer. Yes, I had. I had read them carefully. I knew what needed to change. I knew where the leverage was. The thinking had been done days ago.

The drag came from the manual handoff. Notes in one place. Documents in another. Context scattered across tabs. The kind of friction that doesn't look dramatic but quietly eats your day.

So I opened the beta AI workflow I was building to generate, proof read, format, and collate client documents based on my custom business strategies and frameworks.

On the surface, it was simple. A dedicated client folder and one running strategy document. A trained AI assistant that knew my frameworks, strategies, and my voice.

I dropped in the client documents. The assistant surfaced the relevant sections, pulled my prior notes, summarized the themes, and drafted a clear strategy breakdown in my tone.

I read it once, added one clarifying paragraph and sent it. A moment later, my phone buzzed again.

"This is exactly what I needed," the client wrote. "Thank you for the clarity."

I leaned back and let that land. Because I knew what that moment used to cost me. Fifteen minutes hunting for notes. Seven minutes checking tone. Ten minutes rewriting and second-guessing before hitting send.

Over thirty minutes. Every time.

Back then, those "quick responses" stacked up and stole my evenings. I still remember hearing my kids laughing in the next room while I sat at the table rereading the same paragraph for the fifth time, telling myself I'd join them in just a minute.

This time, it took a few. That was my wake-up call.

I had treated AI like a novelty. I used it for small things but the momentum came when I stopped treating it like a shortcut and started treating it like structure. A structured system that carries repeat work and moves decisions forward without me dragging everything along.

Here's where I'm going to challenge you. Most business owners blame time management. They buy productivity apps, calendars and planners. They reshuffle routines and promise themselves they'll focus tomorrow.

I started to realize that the real problem sits upstream and it lives in interruption and restart.

Studies of entrepreneurs and micro-business owners consistently show burnout driven by wearing too many hats, constant decision making, and the feeling that you can never fully switch off. High stress and mental fatigue aren't exceptions anymore. They're common.

As an entrepreneur, I've worn too many hats, made all the decisions myself, burned the midnight candle and have gotten to the point of burnout.

Working faster feels like control. It gives your nervous system something to hold onto. You stay in motion, checking boxes, solving problems, pushing through. From the outside, it looks productive. On the inside, it's exhausting.

You can get a lot done and still feel completely wrung out.

If your day includes ten interruptions, and even a few of them force a full restart, your best hours are gone before lunch. That's why you feel busy and still end the day unsure what actually moved.

Tools don't fix that on their own.

A Quick Win Before You Move On

Before you read the next chapter, do this.

Open yesterday's calendar, your messages, and your sent folder. Write down the last ten things that pulled your attention. Use plain language. Things like “reply to lead,” “find client document,” or “send onboarding link.”

Next to each one, write a single word: **Judgment** or **Memory**.

Judgment means only you can do it and memory means it just needs to happen consistently.

If five or more tasks land in memory, you’ve just found your first pressure point that needs relief.

Where This Leads Next

The calm morning we started with isn’t a fantasy. It’s the result of building a business that runs through systems instead of constant effort.

This chapter showed why the one-hour workday begins with a different relationship to your time and attention. When that foundation is in place, your business becomes ready for support instead of strain.

But first, you need to see what’s actually running your days.

In the next chapter, you’ll look at your real workday. Not the one you describe to other people. The one you actually live.

Once you see it clearly, the math gets loud.

And when the math gets loud, staying the same starts to feel expensive.

CHAPTER 2

SEEING THE PATTERNS RUNNING YOUR WEEK

Before we go any further, I want you to try a simple test.

Set a timer for the first hour of your day and work exactly the way you normally do. Don't optimize. Don't focus harder. Just move through your morning as usual. When the timer ends, stop and write down every place you went.

For most business owners, a normal morning quietly unfolds like this:

8:00 a.m. – Open email to "quickly check messages"

8:12 a.m. – Click into Direct Messages after seeing a notification

8:25 a.m. – Open a client file to answer a question

8:41 a.m. – Check your calendar to confirm a meeting time

8:55 a.m. – Jump into your notes app to jot something down

9:10 a.m. – Open Google Drive to find a document

9:28 a.m. – Open a new tab to check Stripe after noticing a payment alert

9:45 a.m. – Join a meeting

10:32 a.m. – Exit the meeting and open your CRM to update a record

10:55 a.m. – Open Facebook after seeing a notification badge

11:12 a.m. – Back to email to "clear one thing"

11:35 a.m. – Take a phone call

11:52 a.m. – Check a text message

11:58 a.m. – Back to DMs

By noon, you've touched a dozen tools, restarted the same task multiple times, and made dozens of small decisions without realizing it.

If that list makes you wince a little, that's a good sign. You've just met your real workday.

Most business owners believe they know where their time goes. They truly don't. The day feels full, so it must be productive. But fullness isn't the same as direction.

A Life Coach's Routine

One of my clients, Marco, learned this the hard way. Marco teaches other online coaches how to get certified as life coaches. He runs client sessions, certification cohorts, a weekly podcast, and a paid community. He has one admin, which means the small stuff doesn't disappear. It stacks.

When we first talked, Marco went straight to self-blame. He told me he just needed better discipline. His plan sounded familiar. Wake up earlier, batch content, tighten the community, record more podcasts, try harder. Then he used the word that usually shows up right before burnout does. "Busy." Busy is what happens when your day gets shredded into pieces.

Marco's Wake-Up Call

I gave Marco one assignment. I asked him to track his next fortyeight hours. No cleaning it up later. No summarizing. Just the raw log.

He stared at me like I'd suggested he brush his teeth with hot sauce. "Everything?" he asked. "Everything," I said. "If your finger touches it, write it down or record it in your note taker."

This is what I call The 48-Hour Truth Audit™.

Two days later, he sent me the list. It didn't read like a workday.

It read like a nervous system on duty.

Answering a DM from a lead who said, "quick question", jumping into the community because a student was spiraling, fixing a broken login, replying to a client's voice note, approving a refund, uploading a module, recording a podcast intro twice because the first take felt flat, chasing a Stripe card decline, rewriting the same welcome message so it sounded warm, answering his admin's question about the next live training, starting a sales post, stopping to handle one urgent thing, then starting again. Over and over.

At the end of the log, Marco added one sentence. "I had no idea this is what my day actually looks like."

His weeks were getting busier because his program was growing. Recently, a warm lead went quiet after a delayed reply. A student dropped from the certification program because they felt lost. A sponsor hesitated on a podcast renewal because delivery felt inconsistent.

At home, Marco did what most entrepreneurs do when they swear they're "off". Phone face up, brain running twenty miles and hour, still working, half listening.

When a business runs through your nervous system, you live in constant alert. Remember, that comes with a health cost and a clarity cost.

Volume Isn't the Problem. Fragmentation Is.

Let me say this cleanly. Small businesses carry real volume. Leads, clients, payroll, delivery, tech, marketing, partnerships, and service issues all exist. That part is real.

Fragmentation is what multiplies the weight. Every switch forces your brain to reload, like closing and reopening the same app all day. Research shows that frequent interruptions make work feel heavier and more difficult, even when the total workload hasn't changed.

I used to coin myself the "Queen Of Getting It Done". But what I realized is that working faster is a coping habit. It feels productive for a moment but then it shows up as missed follow-ups, thin patience, missed family time, and that low-grade pressure you carry all day without noticing.

The Lie of the "Quick Task"

Ask a business owner how long something takes, and you'll usually hear the same answer. "Oh, that's quick."

Marco said that about replying to DMs. "It takes just two minutes." He told me. So we mapped the steps.

Finding the message. Scanning context. Checking who the person was. Remembering what was offered last week. Deciding tone. Writing the reply. Rewriting it so it sounded like him. Dropping a link. Checking the link. Setting a reminder to follow up. Forgetting the reminder because something else pinged.

That isn't one task. That's a chain. Chains hide inside "quick" tasks and quietly eat hours.

Putting a Price on Your Time

This is where I push you a little. If you never put a price on your time, every task starts to look like your job.

I use one number for this. I call it your Handoff Number. It's a simple way to see when "busy" starts getting expensive.

Here's how it works.

First, decide what you want to take home this year. Net pay. Not revenue. Revenue lies because it includes expenses, taxes, refunds, and surprises you don't get to keep.

Next, divide that number by 2,000. That's roughly how many hours most people work in a year. This gives you a baseline hourly value.

Then divide that hourly value by 4 to reflect quarterly capacity. That final number is your Handoff Number.

Here's what this looked like for Marco.

Marco wanted to take home $240,000 for the year.

$240,000 ÷ 2,000 hours = $120 per hour.

$120 ÷ 4 = $30 per hour.

That meant every time Marco spent an hour doing work that could be handled for around $30 an hour, he was effectively paying himself $120 an hour to do it.

You wouldn't pay an admin $120 an hour to handle reschedules or card declines, right? But when you don't price your time, that's exactly what happens.

Seeing it this way changes how you look at your precious time right?

The Question That Changed the Audit

After Marco reviewed his log, I asked him one question. "What on this list actually requires your unique judgment?"

He went quiet. Then he circled a few items. Client coaching decisions, student outcomes, curriculum improve-ments, offer strategy, partnership conversations, live sessions.

Almost everything else was glue work. Necessary, responsible, and draining.

The hard truth is that if your business needs you to personally touch everything to keep quality high, the system doesn't have a spine. It has a habit and that habit is hidden as control.

Control feels safe until it costs sleep, trust, and momentum.

Your Forty-Eight-Hour Truth Audit Activity

You can apply the 48 Hour Truth Audit in your own life and business. Just pick one normal hour today, or set aside two normal days if you want the full mirror. Track everything the way Marco did. Ensure its not on a busy week where you are preparing a new offer, a launch, or one where you have an important event to conduct or attend.

Not a crisis week. A regular one.

Write down every task, every open tab, every decision, every switch, and every restart. If writing feels tedious, screen record or leave quick voice notes. This isn't about perfection. It's about pattern recognition.

Log each action with the time started, what you did, the tool used, how long it took, and what triggered it.

When you're done, paste the list into an AI model like ChatGPT using the following prompt:

Act as a lean productivity advisor for solo business owners. Analyze the task log I provide using lean productivity logic to expose hidden effort, wasted motion, revenue leakage, deep focus blocks, and downtime. Group similar tasks even when worded differently and assign each group one primary category only: Revenue Driver, Client Delivery, Content, Operations, Admin, Deep Focus, Downtime, or Learning.

Treat deep focus and downtime as explicit task groups, estimating their frequency, duration, and impact. Count task frequency, estimate total time per task and per category, and show each category as a percentage of total time, including the split between Revenue Drivers and all other work. Flag context switching caused by short, frequent, or interrupt-driven tasks that fracture focus.

Identify the top three low-leverage time drains and recommend whether to batch, automate, delegate, or remove them. Conclude with one focus shift that would protect deep focus and increase revenue leverage without adding hours. Present the results in a clear table using plain language and zero fluff.

You are not asking AI to run your business but instead you are asking it to show you patterns your brain cannot hold.

What to Notice

Take a moment to reflect. What surprised you? What repeated more than you expected? Which tasks drained your energy the most?

What forced constant restarts?

Identify the top three repeat tasks and run each through your Handoff Number. If it doesn't require your judgment and sits at or below that number, it gets handed off.

Even shifting one hour from admin or operations into revenue driving work can change how steady your income feels.

Here's one immediate win. Tomorrow, do not complete the lowest-leverage, non-revenue driving, manual task. Delay it, hand it off to AI or your admin, or remove it entirely and notice the relief.

What to Hold Onto Before the Next Chapter

You can't redesign what you refuse to see. You've just pulled the curtain back on your real week.

A companion workbook for the 48-hour truth audit is available at **www.AIOneHourWorkday.com**

In the next chapter, we'll take what you found and sort it into a structure you can act on. That's where guessing stops, effort gets redirected, and work starts coming off your plate in the right order. That's where time begins to come back.

CHAPTER 3

THE SYSTEMS QUIETLY RUNNING YOUR BUSINESS (AND YOUR NERVOUS SYSTEM)

When your day feels fragile, effort stops helping. You can wake up earlier, plan better, stack habits, or color-code your calendar, and the day will still depend on you noticing, remembering, and catching things in time. The problem isn't discipline. Its design.

Marco didn't come to me asking how to work one hour a day. He asked a much simpler question. "How do I stop feeling behind when I work this hard?"

His interruptions were constant and familiar. A Direct Message (DM) from a lead asking about certification, a student spiraling in the community before a live call, a zoom link that didn't go out, a Stripe payment that failed right before onboarding, a broken Customer Relationship Management(CRM) tag that quietly killed a follow-up, a podcast guest rescheduling two hours before recording. None of it is dramatic and yet, none of it is optional.

Everything pulls at once.

So I told him what comes next. We stop staring at the chaos and we start naming and mapping the engines that are running your business. This will allow us to effectively track the hours you can take back, week by week.

When You Are the System, You Become the Bottleneck

Most owners describe their business as a list of tasks. That language keeps you reactive. Businesses don't actually run on tasks. They run on systems, or what I call "Engines."

Engines are used to create demand for your business or services. They deliver outcomes and they make improvements possible. When an engine has no structure, it borrows your body. Your memory becomes the follow-up system and your attention becomes quality control. Your availability becomes the growth plan. That is loadbearing work I'm sure you are tired of carrying mentally and physically.

This load starts to show up quietly then it gets heavier and heavier over time. We are used to carrying so much context in our head. For example: remembering the unique situations of each client, recalling previous conversations and what needs to happen next, remembering who needs what and by what deadline, etc.,. You find yourself always jumping in so things don't fall apart.

If you don't measure what you're carrying, you keep carrying it. Its not because you choose to, but because the business has nowhere else to put the load. That's why we track it. Not for guilt, but for proof.

The Three Engines Carrying the Weight

Most online and service-based businesses run on three core engines: Growth, Fulfillment, and Innovation. You don't get to opt out of them. If you never design them, they still run. Most of the time they are running inside your head instead of your systems. That's where your exhaustion builds.

Engine One: Growth That Doesn't Depend on You Chasing Leads

Growth is what turns interest into income. For Marco, growth showed up as DMs and sales calls where prospects ask for more details about his certification program, email inquiries, podcast listeners booking calls, referral introductions, and those other "quick question" messages that looked harmless.

Marco thought his growth problem was marketing. It wasn't. His business had plenty of interest and demand. However, his problem was delays.

Delayed messages he meant to answer later, follow-ups that slipped a day, client inquiries that cooled off after a few hours.

In any business, response time matters. Research on lead response consistently shows that the longer a response takes, the lower the chance of conversion, with a steep drop-off after the first few minutes. My take is simple. This isn't about living inside notifications. It's a design issue. A business with structure responds quickly because it doesn't rely on someone remembering.

Ask yourself the following: if ten prospective clients raised their hand today, could your business respond clearly without you stepping in? If the answer is no, growth is still running on personal effort.

Engine Two: Fulfillment That Doesn't Live in Your Head

Fulfillment is everything that happens after someone says yes. It includes onboarding, welcome messages, access to training, certification expectations, live call reminders, community support, progress tracking, and follow-ups when students go quiet.

Marco cared deeply about outcomes. That was never the issue. Instead the issue was consistency. One cohort received clean onboarding, another felt scattered, one VIP student felt supported, and another disappeared quietly.

When fulfillment depends on memory, you stay on alert. You expect something to slip. Eventually, the quiet costs show up. Refund requests. Dropouts. Support messages that feel emotional instead of logistical.

Ask yourself this. If you stepped away for a week, would your clients still feel guided and supported? If that question tightens your chest, fulfillment is still living inside you.

Engine Three: Innovation That Finally Has Room to Exist

This engine determines whether your business improves or simply repeats itself at higher volume.

For Marco, innovation meant improving the certification curriculum, tightening onboarding so students stayed engaged, building clearer accountability systems, refining offers based on real student struggles, and choosing podcast topics that matched actual demand.

He had ideas. He kept pushing them to later. Later never came. Growth and fulfillment consumed the week. When innovation never gets protected space, the business keeps solving the same problems over and over.

A Quiet Realization Before We Go Further

Not every business feels heavy for the same reason. Some feel heavy because they're disorganized. Others feel heavy because the work requires real human judgment.

Structure removes friction. Structure creates consistency. Structure carries weight you were never meant to hold alone. What structure doesn't do is remove the work that actually requires you.

So we redesign where effort lives. Growth can run without you hovering. Fulfillment can run without you remembering. Innovation can run without waiting for a miracle week.

The Number That Turns Fog Into Proof

Marco wanted relief. I wanted evidence. So I gave him one metric:

Handoff Hours.

Handoff Hours are the hours you spend on work that doesn't require your judgment and could be handled at or below your Handoff Number. These hours aren't theoretical. They show up every week. They quietly drain your energy and block progress without announcing themselves.

Research on entrepreneurs and micro-business owners shows burnout is widespread, driven by wearing too many hats, constant decision-making, and the feeling that you're never fully off. High stress and mental fatigue aren't edge cases anymore. They're common.

I agree with the research because I've lived it. Working faster feels like control. It gives your nervous system something to grip. You stay in motion, checking boxes, solving problems, pushing through. From the outside, it looks productive. On the inside, it's exhausting. You can get a lot done and still feel completely wrung out.

Handoff Hours put a number to that pressure. And once something has a number, you can stop arguing with it and start changing it.

Where You Are Now

At this point, you should be seeing your business more clearly. You've tracked where your time actually goes. You've spotted the repeats. You've identified the work that drains you without moving anything forward.

That clarity matters.

Most people rush into AI without understanding their own workload. They grab tools, stack apps, and stay busy while the real problem remains untouched. You didn't do that.

You now know what needs support. You know what should stay human but you also know that the right structure will give you back your time.

The next step is knowing who builds your support structure.

AI isn’t one thing. Writing, creating, storing, processing, and running automated workflows require different capabilities that are unique to each tool. When people treat AI like a single tool, results fall apart.

In the next chapter, I’ll walk you through the core types of AI you’ll actually use in a real business. I'm anti-AI-overwhelm so we will take it slow. Just practical clarity so that AI stops feeling overwhelming and starts feeling useful.

PART II

BUILDING THE AI BACKBONE THAT TAKES THE WEIGHT OFF YOU

CHAPTER 4

THE THREE TYPES OF AI THAT CHANGE EVERYTHING

I love the fact that in today's digital era, you don't need to be technical to lead a digital team. You just need to understand who holds what role. I spoke about this in a recent podcast episode. You can find it on

www.TheAIOneHourWorkday.com

Most people feel overwhelmed by Generative AI because they first encounter it the wrong way. They open a tool, stare at a blinking cursor, and assume they're supposed to magically know what to ask. Or they jump into a course promising to get AI to sound exactly like them. When the output feels generic or doesn't work the way they imagined, they decide AI doesn't work, or worse, that they're bad at it.

However, that's not the real problem.

The real problem is that no one explained the funda-mentals. No one explained the roles AI plays or what each type is actually good at.

This book is not a technical deep dive into models or theory. You can learn more about AI at

www.TheAIOneHourWorkday.com But you do need a clear understanding of the three types of AI most business owners already use, often without realizing it, because these are the tools that will eventually make your AI-Powered One Hour Workday possible.

Think of it like walking into an office where three specialists, with different roles, are already working. One thinks and writes. One creates visuals assets. One runs tasks exactly the way you would if you had the time and attention. They don't get tired.

They don't forget. They don't need constant supervision. They aren't people. They're the three types of Generative AI you already have access to. Once you understand their roles, AI stops feeling abstract and starts feeling practical.

The Brain and Writer: Large Language Models (LLMs)

The first member of your AI team is the Thinker and Writer. I call it "The Brain."

This type of AI is trained primarily on text. At its core, it predicts what comes next in a sequence of words based on patterns in the human language. If you say, "The sky is," it predicts the next word should be "blue." That simple prediction engine is what allows language models to write, summarize, organize, and reason through information.

Tools like ChatGPT, Gemini, Claude and many other LLM models fall into this category. They're especially good at working through large amounts of information, the size of entire encyclopedias, and turning it into something usable. Research from institutions like Stanford has shown that modern language models excel at text generation, summarization, and reasoning, which is why they fit so naturally into everyday business work.

In real life, this means your monthly planning and email calendar is done before your coffee gets cold. One prompt. A full month of campaigns drafted in your voice, with subject lines, call to actions, and angles that actually match what you sell. No blank screen. No "I'll come back to this later."

It means walking into your next content cycle with podcast titles, outlines, and scripts already mapped from what's trending in your market right now. Not generic ideas or recycled guru takes. Your own unique positioning in your language. Built to pull the right people into your world.

It means your brain stops being the storage unit for half-finished ideas. Meetings get summarized and decisions get captured and routed. Messy thoughts get turned into clear next steps you can actually execute. Nothing important falls through the cracks because it's all written down, organized, and ready to move.

The bottom line is, you stop spending your best hours trying to "figure out what to do," and start using them to build the things that grow revenue, leverage, and freedom.

That's what this looks like when AI is working for you, not you working your precious human hours for it.

I remember the first time I used a language model for something real. I asked it to help me outline a piece of content I had been avoiding all week. The outline came back in seconds. It wasn't perfect, but it was clearer than anything I had been circling in my own head.

That's when I realized this wasn't about replacing my voice. It was about getting me out of my own way.

A language model helps you to move faster once you have a rough idea of what you want to say or do. It's especially powerful when you're explaining something you've explained a hundred times before. Creativity no longer depends on the right mood, the right time, or a perfectly quiet room. Instead of asking, "Can I think of something?" you start asking, "Which version feels most like me?" That's a very different place to create from.

The Visual Creator: Multimodal AI

The second member of your AI team is the Creator, often called multimodal AI. This type of AI understands more than text. It understands the relationship between images, audio, video, documents, and the words used to describe them. It can work with screenshots, PDFs, voice notes, images, videos, and mixed inputs at the same time. You'll find multimodal AI embedded in many tools today, including ChatGPT.

Researchers at organizations like Google DeepMind have shown that multimodal systems improve AI model comprehension by combining different types of information. This type of AI training helps the models to improve daily and will soon lead to Artificial General Intelligence (AGI) which is the next anticipated level of AI that will be able to perform any intellectual task a human can.

In practice, multimodal looks like creating visual concepts for presentations, dropping in screenshots and asking what's missing, turning voice notes into diagrams, or generating images and assets without waiting on a designer.

For me, multimodal AI changed the creative side of my business completely. I use it to generate headshots, graphics, PDF and PowerPoint presentations, AI clones, videos, and logos. What used to take weeks of design, prototypes and coordination now takes minutes. Instead of waiting for final drafts, I can create what I need on demand and keep moving forward.

The Operator: Agentic AI

The third member of your AI team is the Operator. This is usually the one people meet last, and it's the one that changes everything.

Agentic AI is built by connecting a language model (LLM) to memory, rules, and actions. You'll find versions of it inside tools like ChatGPT, Claude, Google AI Studio, n8n, Go High Level, and Notion, just to name a few. I love that agents don't just respond but they take action.

They can run multi-step processes, follow schedules, and execute tasks based on your rules without needing you to prompt them every time. Unlike a stand alone large language model, an agent doesn't have to wait for you to start it each time. Once it is programmed, it runs autonomously in the background, following the frameworks, schedule, decisions, and boundaries you've already set.

I won't get into the technical setup for agents here. What matters is this, AI agents take over the manual work you've been carrying for years, giving you much needed time back.

I see too many business owners drained by mental handoffs and repetitive tasks, when a single well-set AI agent could give them hours back every day and move them closer to a real one-hour workday. It still amazes me when I watch an agent take a client's data, run competitive analysis, generate weekly financial comparisons, and surface clear strategies and insights. That same work would have taken a human analyst four full days.

Research from organizations like IBM notes that Agents increase reliability because they execute repeatable steps the same way every time. Nothing slips. Nothing gets skipped.

In practice, this can also look like pulling leads from the web into a database, responding to new inquiries automatically by text, phone, or email, sending onboarding messages the moment someone pays, summarizing calls and sending follow-ups, building simple websites or ads, or even handling inbound calls as a voice agent.

The point isn't novelty, it's relief. When AI agents carry out the repeat work, your energy goes back to decisions, direction, and the parts of the business that actually need you.

I still remember the first task I handed to an agent. It felt uncomfortable, like I should be hovering over it in case something went sideways. That's normal. When you've been the safety net for years, letting go feels risky at first.

One of the most beginner-friendly agents I recommend starting with is the ChatGPT agent. It's simple to set up, it asks for permission before acting, and it can't touch your tools unless you explicitly allow it. That's true of any agent you'll use responsibly.

Nothing happens in the background without your say.

One of my favorite moments came when I watched an agent take a messy list of tasks and turn it into a clean weekly calendar inside Notion. Then I asked it to organize more than two hundred prompts into a structured database. It built everything while I went downstairs to make a smoothie and a sandwich.

That's when it really landed for me how much time and energy I'd been burning on manual work that never needed my attention.

So here's the challenge I give you. Anytime you feel yourself slipping into a manual task rabbit hole, trigger one simple cue in your head. One Hour Workday. Let that pull you back into alignment with the business you're building.

Pause and ask yourself why you're doing this by hand instead of letting an agent handle the grunt work. Your time is worth more than busywork. Hand it off. Use that reclaimed time to book the massage you've been postponing or reconnect with a family member you haven't talked to in too long because you were always "too busy."

When the Team Works Together

On their own, each type of AI is helpful. Together, they change how work happens. Harvard Business Review has pointed out that the strongest productivity gains come when automation systems are coordinated instead of scattered.

When your AI team works together, the writer drafts, the creator multiplies, the processor organizes, and the operator executes. You effectively gain a writing studio, a creative department, an operations assistant, and a process executor, all supporting you.

None of them replaces you.

Why This Matters for the One Hour Workday

The one-hour workday is not about doing nothing. It's about not doing everything.

Once you understand these roles, you stop asking AI to do random things and start assigning it real work. Writing. Creating. Processing. Executing. You stop relying on raw effort and start relying on a team.

Now that you know which types of AI specialists can be plugged into your business, the next step is deciding where they help first.

In the next chapter, we'll map your weekly workload, spot repeating patterns, and identify the tasks AI can take off your plate immediately. That's where the one-hour workday stops being an idea and starts taking shape.

CHAPTER 5

BUILD YOUR AI BRAIN BEFORE YOU AUTOMATE ANYTHING

Before you build a single automation, I want you to stop and answer one question: "Do I want AI to guess, or do I want it to be 100% accurate?"

Marco learned this lesson quickly. When he first came to me, he was excited about AI for content. Faceless videos, captions, AI clones, promotion ideas. He jumped in and tried it.

“It sounds fine,” he told me. “It just doesn’t sound like me.”

That was my clue. I knew where we could begin with my AI One Hour Workday Operating System.

AI can’t support your business if it doesn't know your writing style, your voice, or has no memory of how your business works. Without context, it fills in the gaps the only way it can: by guessing. So we didn’t start with content, agents or automations. We started by building a brain.

The Only Rule You Need Before Using AI

I gave Marco one simple filter that kept him grounded and kept his business human. This is my HAS Framework™: Human Only, AI Automated, and Supported by AI.

If a task requires your judgment, it stays **human.** If a task is manual, repeatable, and predictable, it should be **automated** by AI. If a task needs your voice or taste but not your full attention, AI **supports** it by drafting and organizing so that you can review the draft output and use your judgement on how to proceed.

This single rule prevents two common breakdowns. People hand AI decisions it should never make, which breaks trust. And people cling to repetitive work because it feels small, even though it multiplies quietly.

What a Knowledge Base Actually Is

A knowledge base is memory stored inside an AI language model. In tools like ChatGPT, this shows up as a Custom GPT where you upload files as reference material. Those files become the foundational context that the model uses instead of guessing.

My view is practical, AI doesn't need your entire Google Drive. It needs clarity. Which means your frameworks, your standards, your offers, your tone, your rules.

Once that memory exists, AI stops sounding generic. It starts sounding like you.

This is why we didn't build one massive brain for Marco. One giant knowledge dump turns into a junk drawer. Focused brains create leverage. So within the AI One Hour Workday Operating System, we built five, each with a specific job.

Marco's Five AI Brains

The first was his program brain. This one learned his life coaching certification program from the inside out. We fed the brain his signature method, program structure, course notes, workshop materials, success metrics, and examples of what he promises and what he never does.

He used it to draft lesson summaries in his voice, generate milestone-based student feedback, create weekly focus guidance, and turn messy notes into clear coaching points. Before this, he spent around thirty minutes a day re-explaining concepts. Afterward, he spent about five minutes reviewing drafts. That alone saved him roughly two hours a week.

The second brain handled client operations. This one focused on onboarding and fulfillment. It knew the steps, the order, the tone, the escalation rules, and the standards. Marco used it to draft onboarding emails instantly, answer process questions without searching, keep his tone steady even when tired, and give his assistant clear scripts. That cut his weekly time from over an hour to about twenty-five minutes.

The third brain was built for his virtual assistant. This one reduced interruptions. It held rules for what could be handled independently, what needed escalation, what never reached Marco, and how responses should sound. His assistant used it daily.

Marco stopped getting random pings and started receiving clean summaries. What used to take him an hour a week dropped to about ten minutes.

The fourth brain was his brand voice. This is the one that solved the "sounds fine but not like me" problem. We uploaded welcome emails, nurture sequences, podcast scripts, captions, and even voice notes of Marco speaking naturally.

He used it to draft emails that actually sounded like him, turn voice notes into clean drafts, and move faster because the starting point already fit. His weekly content time dropped from three or four hours to about ninety minutes.

The fifth brain handled FAQs and operations. This one removed daily repetition. It knew pricing explanations, policy language, certification steps, and access instructions. Marco used it to reply to DMs quickly, give approved answers to students and his assistant, and close mental loops. What used to take nearly an hour a week dropped to about ten minutes.

What Marco Gained Before Any Automation

Before we touched agents or workflows, Marco reclaimed time. Realtime. Roughly six and a half hours a week, simply by giving AI a brain so it could stop guessing.

That's the point. You build the brain first. Then automation becomes safe.

A Simple DIY Way to Start

This isn't homework but instead it is guidance. Start with your 48-hour truth audit and label a few tasks using the HAS framework. Choose one decision that stays human, one repeated question that should be automated, and one thing you draft often that AI could support.

Create one Custom GPT and name it for the job. Upload five to ten strong files, such as a core framework, an onboarding process, a support guide, real FAQs, or emails written in your voice, or your voice notes so the way you uniquely speak is captured.

Then test it with real prompts.

Prompt example: "Act as a seasoned business coach. Draft a short, confident reply to a brand-new student who asks, "Where should I start?" Use a calm, supportive tone that sounds like a real human, not a guru. Keep it brief, clear, and reassuring. Focus on reducing overwhelm and pointing them to the single best first step. Write in first person. No fluff. No jargon. No emojis."

Track one metric as you go. Minutes saved today. If that number doesn't drop, improve the inputs before adding more tools.

What Comes Next

Before automation, Marco now had something most founders never built. A business brain that lived outside his head.

That's when we moved into agents.

An AI agent without a properly trained brain can create chaos, but an AI agent trained with the right context and framework will create relief.

In the next chapter, you'll see how I helped Marco connect these knowledge bases to his first agents, so lead response, sales, onboarding, and follow-ups ran without him being tethered to his phone. That's when the one-hour workday starts to feel real.

CHAPTER 6

WHEN THE BRAIN TAKES OVER THE BUSYWORK

Marco didn't come back to our next call excited. He came back steady.

He didn't talk about any recent mishaps, firedrills, or missed metrics. He said, "I finally ate lunch without checking my phone."

We chatted some more and I began to see more silent relief before he even realized them. the signal. He was able to quantify how his new systems design was changing the way he ran his business and best of all, his life.

Up to this point, Marco hadn't automated chaos. He had done something quieter and far more effective. He moved memory out of his head and into systems that could hold it without distortion. Chapter 5 built the brain. This chapter shows what happens when that brain starts doing the work.

Why Automation Finally Worked This Time

Here's what most people get wrong about automation. They treat it like a starting line when it's actually a finishing move.

Automation only works once decisions are already made. When people try to automate motion before clarifying judgment, they end up with brittle systems that feel cold, wrong, or constantly in need of tweaking.

Marco avoided that trap. By the time we introduced agents, his frameworks were written clearly, his standards were explicit, his tone was trained, his processes were documented, and his boundaries were defined.

So when agents entered the picture, they didn't invent behavior. They followed the framework that I installed for Marco. That's the difference between automation that creates relief and automation that creates chaos.

The Same HAS Filter Still Applies

Nothing new was added here. The same filter stayed in place.

Decisions, values, acceptance criteria, and direction stayed human. Routing, responding, reminding, and tracking became automated. Drafting, summarizing, and preparing context stayed supported by AI. The agents didn't blur those lines but they instead enforced them.

Agent One: Lead Response Without Urgency

We started where pressure entered the system. The front door.

Marco's leads arrived through intake forms, direct messages, his podcast, ads, his website, and referrals. Each one created a small spike of urgency. Did I reply? Should I reply now? Will this turn into a long back-and-forth?

We built his first agent using three existing knowledge bases: brand voice, client operations, and FAQs and policies.

The agent responded within minutes, routed every inquiry to a single intake path, asked a small set of qualifying questions, set expectations clearly, and flagged qualified leads for Marco.

Before this, each inquiry cost him several minutes. Across an average week, that added up to about ninety minutes of scattered attention. Afterward, he received one daily summary and spent about ten minutes reviewing it. The time saved was noticeable, but the emotional shift mattered more. His phone stopped feeling urgent.

Agent Two: Onboarding Without Anxiety

Once the lead response stabilized, the next pressure point showed up. Onboarding.

The same questions kept repeating. The same confusion surfaced. The same follow-ups landed in his inbox. This agent ran on his client operations and FAQ knowledge bases.

It sent welcome messages immediately, delivered clear first steps, directed clients to a single home base, and answered common onboarding questions. Marco stayed present for real conversations, and orientation stopped depending on his memory.

What used to take him about an hour a week dropped to handling only true exceptions. Confidence increased. Apologies disappeared.

Agent Three: Calls Without the Scramble

The calls themselves were never the problem. The scramble around them was.

Before each call, Marco searched for notes, rebuilt context, and worried about forgetting what had already been decided. After calls, he spent time summarizing and sending follow-ups.

This agent used the program brain and client operations knowledge bases to pull client history before calls, summarize current status, draft post-call recaps, and list next steps. Marco showed up focused, and memory stopped being a requirement.

What used to take ten minutes per call dropped to a quick review. Over the course of a week, that reclaimed nearly forty-five minutes and removed the mental residue that lingered afterward.

What Changed in a Real Week

From just these three agents, Marco reclaimed time without adding effort. Roughly forty minutes from lead response, fifty minutes from onboarding, and forty-five minutes from call preparation and followup.

When added to the time he had already reclaimed by building his knowledge bases, Marco was no longer operating in reaction mode.

In total, he took back about eight and a half hours each week.

No theatrics. No heroic push. Just fewer things pulling at him.

Why This Didn't Feel Like Automation

Marco never said he felt faster. He said he felt quiet.

"I'm not wondering who I forgot."

"I'm not checking my phone between sentences."

"I'm not carrying it."

That's containment and that's what happens when loops are closed.

The Control Rule Still Holds

AI handled repeat work, routing, summarizing, and preparation.

Marco handled decisions, exceptions, values, and direction.

Any time AI drifted, we corrected the inputs. Control stayed human.

Why You Felt the Difference First

Understanding without relief doesn't change behavior. That's why this book lets you feel the shift before explaining the mechanics.

Pressure drops. Clarity shows up. Better questions follow. You're not here to run an AI lab. You're here to protect your time and attention.

What Comes Next

Once busywork is contained, a new question surfaces. Where should AI never be used?

Some moments require presence. Some moments require judgment. Some moments break if you rush them.

The next chapter draws that line clearly. That's how the AI-Powered One Hour Workday stays steady instead of brittle.

CHAPTER 7

A SPA OWNER'S COST OF BEING THE DECISION ENGINE

Have you ever asked yourself: "If I disappeared for one day, will my business lose customers or reputation?"

My client, Elena, knows the answer. There would be significant impacts because her staff relied on her for decisions each day.

Elena owns multiple spa locations. On paper, things look solid. Her calendars are full. Clients return regularly. The brand is recognizable.

From the outside, it looks like a well-run operation.

In practice, her days dissolve into decisions. Not big strategic ones, but small operational ones that arrive nonstop. Her workday doesn't break because she isn't trying hard enough but it breaks because every system still routes uncertainty back to her. You will see the difference between running a business and carrying one.

The Three Moments That Explain Her Business

You can understand Elena's business by watching three moments play out.

The first is the double booking. She's driving between locations when a text comes in. "Room three is double-booked. What should we do?" Elena replies quickly, like she always does. Who to move.

What to comp. How to explain it. The decision itself takes seconds.

The interruption costs far more.

The front desk didn't escalate because they were incapable. In fact, they escalated because the system never learned how to decide without Elena driving it.

The second moment is the no-show spiral. A therapist doesn't arrive. A client is waiting. The front desk panics. Elena gets pulled in. She scrambles coverage, issues a refund, and loses trust she won't fully recover. No-shows don't hurt first at the register. They hurt confidence. And confidence erodes quietly.

The third moment is the one-star review. It's late, and Elena is writing the response herself. The client felt ignored after a scheduling issue days earlier. Slow follow-up. Confusion. Frustration. Elena can trace the real cause immediately. A missed call. A delayed response. A breakdown that never looked urgent at the time.

Reviews don't start online but instead, they start in your operations.

The Belief That Keeps Owners Stuck

Most multi-location owners carry the same belief. "I have to approve everything so quality stays high."

Here's the reality Elena had to confront. If quality only exists when you personally touch every decision, then quality is already unstable. It just happens to have your name on it.

That belief doesn't protect standards. It hides weak systems.

Why Coordination Breaks Owners

Elena isn't drowning in tasks. She's drowning in handoffs.

Appointment changes. Staff call-outs. Front desk ques-tions. Billing issues. Inventory gaps. Training drift. Reputa-tion fires. Each issue asks the same question. Can you decide this?

And because Elena cares, the answer stays yes.

The Engines, Rebuilt for Spa Reality

The AI One Hour Workday Operating System still runs on engines.

They may just look a bit different in various business environments.

The growth engine shows up through calls from Google listings, website bookings, gift card purchases, concierge referrals, walk-ins, and rebooking at checkout. The failure point isn't demand. It's speed and follow-up. Missed calls. Slow replies. No rebooking prompts. No recovery after no-shows. When response lags, revenue leaks quietly.

The fulfillment engine includes booking confirmations, prep instructions, intake forms, membership status, room readiness, therapist notes, post-service follow-up, and issue resolution. The failure point here is memory. Details live in people's heads and text threads. Quality varies by location, and clients feel the inconsistency immediately.

The innovation engine is where improvement should happen. Reducing repeat complaints. Raising rebooking rates. Improving membership retention. Training staff to upsell cleanly. Lowering noshows with better deposit structures. The failure point isn't ideas. It's time. Elena never gets space to analyze patterns because she's always reacting. Innovation turns into a wish.

The Fourth Engine Multi-Location Owners Need

Solo operators can sometimes ignore this. Elena can't.

Multi-location businesses run through people and consistency. If staff can't execute the system without escalation, the owner gets pulled back in. That's when growth stalls, bad reviews spike, and burnout accelerates.

Where AI Fits Without Breaking Trust

This is where skepticism usually shows up. Elena didn't want robotic replies or cold automation. She wanted fewer escalations.

Language models helped standardize thinking. Front desk responses became clearer. Complaint handling followed consistent rules. Training was explained once instead of repeatedly. Expectations were written down instead of remembered.

The AI One Hour Workday Operating System(OS) agents enforced execution so that any missed calls would automatically trigger an automated follow-up. Confirmations and reminders ran automatically. Cancellation rules were applied evenly. Issues were routed before they reached Elena.

AI didn't replace Elena's judgment but protected it instead.

What Changed First

The first change wasn't dramatic. Elena stopped getting texts while driving. The front desk resolved double bookings using default rules. No-shows followed a consistent response. Clients felt cared for without escalation. The business got quieter.

Quieter = capacity.

Why This Holds Up Under Pressure

Research highlighted in Harvard Business Review shows that operational consistency matters more to customer trust than individual heroics.

My take is simple. Clients don't remember effort. They remember how stable things felt. Consistency beats availability every time.

Why This Chapter Matters

Elena's overwhelm was not because she lacked discipline but because her business trained her to be the decision engine.

The AI One Hour Workday OS doesn't remove owners from their businesses. It removes the business from their nervous systems.

In the next chapter, we'll map Elena's exact pressure points and show how to apply these same principles whether you run one spa or ten. Relief that only works during calm weeks isn't relief.

We're building systems that hold when things get busy.

CHAPTER 8

TEACHING THE BUSINESS TO ANSWER WITHOUT YOU

If your team stopped messaging you for one day, would the business run more smoothly, or would it freeze?

Elena already knew the answer.

Her spas didn't slow down because she lacked hustle but instead, they slowed down because every gray area climbed the ladder until it landed on her. Unclear rules, fuzzy boundaries, and undocumented decisions turned her into the help desk for everything.

So we did one thing first. We stopped treating Elena like support.

Fixing Routing Before Adding Tools

I told Elena we weren't adding new tools. We were fixing routing. We ran her daily chaos through the same HAS framework you've already seen. Anything that carried real consequence stayed human. Refund exceptions beyond policy, hiring and termination decisions, sensitive client escalations, and menu changes remained with her.

Anything repeatable ran without her. Confirmations, reminders, emails, missed-call follow-ups, late cancellation rules, membership billing explanations, and inventory triggers moved into automation.

Anything that needed her voice but not her full attention became supported by AI. Complaint replies, front desk scripts, training explanations, and escalation summaries were drafted for review instead of pulled from her in real time.

This step reduced stress fast because it removed ambiguity. Teams escalate when decision boundaries feel unclear. Research highlighted by Harvard Business School shows that clarity around decision rights improves execution under pressure. I've seen this firsthand. When rules live in one person's head, the entire organization becomes a waiting room.

Building a Shared Brain for Staff

Inside the AI One Hour Workday OS Elena's first Custom GPT wasn't for marketing. It was for operations.

We built a Spa Operations GPT used by front desk staff, managers, and shift leads. Its job was simple, answer questions the way Elena would, using Elena's rules.

We loaded it with front desk procedures, booking rules, deposit and cancellation policies, refund thresholds, membership billing scripts, service prep standards, tone guidelines, and examples of approved responses.

The staff loved using the GPT instead of feeling guilty about bugging Elena.

They asked the GPT questions like:

"What's the default fix for a double booking?"

"What do we say when a client wants a refund for a late cancellation?"

"The therapist called out. What's the coverage playbook?"

"A membership payment failed. What's the script?"

That single GPT turned Elena's standards into a shared brain.

Before this, she handled fifteen to twenty interruptions a day, each taking a couple of minutes. That added up to thirty or forty minutes daily. Afterward, only true exceptions reached her. About five minutes a day.

The time savings mattered, but the bigger win was focus. She stopped getting pulled out of real work for micro-decisions.

Automating the Front Door

Most service based clients are not casual online inquiries. They don't just fill out a form and wait. They call. They try to book, and if they don't get an answer, they call the competition across town. These leads need to be captured immediately so that the business stops missing sales.

For Elena, the same problem showed up every week. Missed calls and slow follow-up. Sales research consistently shows that when response times increase, revenue drops. I've seen this play out again and again. Business owners invest money in ads to bring in leads, but without automated responses or enough support in place, those leads slip through the cracks. The customer doesn't wait. They call the next spa, the one with a chatbot or booking system that can respond immediately and lock in the appointment.

So we automated the front door. Missed calls triggered an AI agent that sent a simple text inviting the client to book. Booking confirmations included clear prep instructions to eliminate back-andforth questions. Deposit and cancellation policies were explained upfront, so staff don't have to argue with the customer.

Before this, Elena spent about forty-five minutes a day putting out booking fires. Afterward, she reviewed one short daily summary in about ten minutes. Just that one shift gave her back roughly three hours a week and quieted the constant urgency coming from her phone.

Why This Matters More Than Marketing

I've worked with many entrepreneurs who are doing the "right" thing by spending money on advertising, but still losing money because their business isn't set up to catch what the ads bring in.

The leads are real. The interest is there. But without systems to capture, respond, follow up, and guide people to the next step, the opportunity disappears.

That's where the frustration comes from. The marketing works, but the business can't keep up. Automating the front door isn't about being fancy. It's about making sure the effort you're already paying for doesn't leak out the sides.

Tightening the Client Journey

This is where spas quietly bleed time. When systems feel scattered, clients feel ignored. Staff improvises, quality differs from location to location and complaints rise.

We set one default path for every client, from confirmation to arrival prep, check-in scripts, service notes, post-visit follow-up, and rebooking prompts. Elena stayed involved in moments that required presence. The system handled the predictable ones.

What used to take about thirty minutes a day fixing preventable issues dropped to handling only real exceptions. That reclaimed another two hours each week.

Handling Reviews Without Using the Owner as a Safety Net

Elena had been writing review responses late at night. That's a sign the business is still using the owner as the last line of defense.

We built a simple review response agent with one rule. Acknowledge the issue, apologize when appropriate, and invite the conversation offline.

Research on service recovery shows that consistent, calm public responses protect trust, while private follow-up resolves the relationship. The public reply protects your reputation. The private conversation fixes the problem.

What used to take Elena nearly forty minutes a week dropped to reviewing a short summary in about five minutes.

Cutting Noise with a Website Chatbot

Finally, we added a website chatbot to handle questions that quietly steal energy. Recurring customer questions about pricing, hours, membership details, prep, parking instructions, and cancellation policies no longer landed in her inbox or at the front desk.

That alone removed about forty minutes of repeated questions each week. The AI One Hour Workday Operating System changed Elena's business and even her outlook on the capabilities of AI.

What Elena Reclaimed

When we added it all up, Elena reclaimed nearly nine hours a week. That's a full workday returned. No grind speeches. No complicated dashboards but instead, fewer things climbing the ladder to her.

What Comes Next

Elena did not get her time back by working faster but by removing noise.

Most days now include one protected hour. No more front desk pings, no more recurring policy questions, no more bad customer complaint fires. She uses that hour to think, plan, strategize and design offers that her competitors haven't noticed yet.

We keep testing the system under pressure. Busy weeks, staff call-outs, fully booked schedules. Clusters of complaints. Each time, we return to the same question. Does this still protect her "one hour?"

Because a one-hour workday only counts if it holds when things get messy. That's where most systems fail. They route cleanly until a gray area appears, and everything climbs back to the owner.

The next chapter solves that. It's where we stop asking systems to guess and start teaching them how to decide. Because the moment your business needs your judgment for everything again, relief disappears.

If your systems can't make the right call without you, you don't have leverage yet. You have automation with training wheels.

Next, we teach your systems how to decide while keeping control exactly where it belongs.

CHAPTER 9

HOW TO TEACH YOUR SYSTEMS TO MAKE THE RIGHT DECISIONS

Stop checking the logs.

If your business finally got quiet but your brain is still on watch, you're normal. You're just not used to the quiet instead of the chaos.

Marco ran straight into this after his first real stretch of relief. His phone stopped buzzing and his lead intake and replay system ran on its own. Onboarding stopped dragging out for days. And then he said, almost sheepishly, "I keep checking the logs. I just want to make sure it's doing the right thing."

That sentence is the doorway into this chapter. Sometimes it takes time to trust that your AI software is doing the right thing.

Why Systems Break After They "Work"

Most systems perform just fine with clean cases but they break when you introduce edge cases.

Let's say a one-off request happens. Something outside of your AI instructions. For example, a client sends three paragraphs of emotion with one actual question buried in the middle. This could lead to a messy situation. When that happens, founders panic. They pull everything back into "I'll do this myself" and decide automation is unreliable.

The real problem is simple. Your system rules were not clear enough to tell the AI what to do with edge cases.

Humans carry rules in their heads. Systems can't. They need those rules written down.

Research on agent-based systems frames this clearly. Agents succeed when leaders define boundaries, roles, and oversight so execution stays inside guardrails instead of drifting.

It's pretty simple to me. Business owners don't only need a smarter system but a clearer one.

The HAS Filter Still Runs the Show

Keep the HAS framework and the AI one hour workday cue at the forefront at all times. This will help you to stay on track when edge cases show up, or those shiny new tools, or context switches creep up.

Remember, if a decision carries real consequence, it should stay human. If it's repeatable and rule-based, the system runs it. If it needs your voice, your input, your oversight, AI prepares the first draft, and then you polish it up with your own perspective. This filter keeps your business human without turning you into the bottleneck.

The Three Decision Types Every Business Has

Every decision in your business falls into one of three buckets.

Fixed decisions like pricing rules, refund policies, response windows, and eligibility criteria can be handed off to AI. If you redecide these every time, you burn judgment on routine.

Pattern decisions are choices you make over and over again. For example, is this lead a good fit? Does this need attention right now or can it wait? Does the client need information or a call?

Your AI-powered system will decide, segment, route and handle these inquiries without your direct involvement. Verizon does it everyday when you call in and speak with their AI voice agent.

Now you can use a similar system at a much lower cost for your business. You are the architect. You are still in control of how your AI systems will be custom designed for your business.

As the Human in the loop, judgment decisions stay with you. You will need your own human judgment for conflict resolution, ethical lines, and strategic direction. If a decision involves trust or long-term outcomes, it remains human.

Once Marco sorted his decisions this way, his anxiety dropped.

Nothing important left his hands. Only repetition did.

How to Teach a System to Decide Without Guessing

Don't start with prompts but instead, start with one decision you already make every week.

The forty-eight-hour truth audit shows you where these decisions hide. You'll see the same question showing up again and again, just wearing different outfits.

Marco chose one he answered constantly. Is this new coaching student ready for certification right now?

He was answering it five to ten times a week using the same mental steps. So we pulled it out of his head.

First, we wrote the decision as a short checklist. The qualification criteria for his new coaching students were: they must have prior coaching experience, ability to commit to live calls, clear intent to use certification professionally, and the ability to start within thirty days. This wasn't new logic but was already how Marco decided. We just made it visible.

Next, we assigned outcomes to each path. If all conditions matched, the system booked a consult. If some matched, it sent prep materials and invited follow-up. If none matched, it sent a waitlist resource and closed the loop. No guessing or improvising.

Just routing.

Finally, we required reporting every time. A system that acts silently keeps your nervous system on edge. So for every new coaching student, we generated a short summary and a clear tag: student qualified or, needs further review, or not a fit.

Marco now reviewed the qualification summaries once a day and stopped hovering all day waiting for a new student registration email to come in.

What This Saved in Real Terms

Before this change, Marco spent about twelve minutes per lead reading messages, checking context, thinking through fit, choosing next steps, and writing replies. With eight leads a week, that added up to over an hour and a half.

After the decision logic lived in the system, he reviewed a single batch summary. About ten minutes.

The time Marco got back mattered and so did his psychological relief, which mattered even more.

That dream of mental relief is not far away for you.

Why This Calms Your Body

Uncertainty keeps people in a heightened state while they wait for clarity. Psychological research consistently shows that ambiguity drives stress and vigilance, especially when outcomes feel important and unresolved.

My view is simple. When your system guesses, your body braces and when your system follows clear rules, your body rests.

That's what trust feels like.

The Line AI Never Crosses

Marco kept certain decisions firmly human. Client acceptance and rejection, conflict handling, and refunds. He enforced boundaries so nothing would get out of control. More important, he ensured his business vision, his human touch, and his strategic input remained intact.

That balance protected trust while the system carried the repetitive work.

A Quiet Test You Can Run This Week

No worksheets required.

Notice how many times you answer the same decision question.

Not the task but the decision.

That's your next system target. The mental offload.

You're not handing away control but instead, you are cleaning it up.

Where This Takes Us Next

At this point, your systems respond, route, and decide inside clear boundaries.

The last problem is subtler. A system can be correct, and the business can still feel heavy. This is either a design issue or a mental load issue that needs to be cared for immediately.

In the next chapter, we design the workday itself so it holds during launches, business spikes, and messy weeks. That's how the One Hour Workday stays steady when things get real.

PART III

MAKING THE AI ONE HOUR WORKDAY HOLD IN REAL LIFE

CHAPTER 10

DESIGNING A WORKDAY THAT HOLDS WHEN LIFE GETS LOUD

If you had one quiet morning last week and then filled it with more work, pause. That reflex comes from being the DIY person, instead of the strategist.

Marco did it the first week his systems started working. Elena did it once her front desks stopped lighting up her phone. The day finally got quiet, and they treated the quiet like a problem to solve.

That habit will swallow your time again, even with great systems.

Why Your Day Refills So Fast

This is what it looks like in real life.

Marco finished his protected hour early, saw a quiet inbox, and started tinkering. He tweaked a landing page that was already converting. He answered a "quick" DM that could have waited. He reviewed team work that had already been approved.

Elena saw a calm afternoon between locations and jumped back into the weeds. She checked the staff schedule "just to be sure." She reviewed inventory reports she already automated. She personally responded to a review her system had already flagged.

Nothing exploded. The day just stretched back out.

This pattern has a name. Parkinson's Law describes how work expands to fill the time available for its comple-tion. If you don't set a stop point, work will keep volunteering itself.

What the One Hour Workday Means in Practice

The one hour workday works when you treat your hour like a decision studio.

Marco used his hour for decisions that moved outcomes. He decided which certification candidates to accept. He tightened the curriculum based on what students struggled with. He recorded teaching content with no interruptions.

Elena used her hour to lead, not manage. She reviewed location performance summaries, made pricing and membership decisions, and planned seasonal offers across locations.

They did not use that hour for email, scheduling, or follow-ups.

Those were already handled by systems.

The Rule That Protects the Hour

During your protected one hour workday, the business can't pull you back into motion.

No texts, no inbox checking, no booking software, no staff interruptions.

If something doesn't require judgment, it waits. If something requires unique judgment, it gets slotted for the hour.

Owners break this by letting one small thing slip in. The one message, one quick check, one "real fast" reply. That opens the floodgate again.

How the Rest of the Day Runs

The day stays calm when information flows without chasing.

Before the hour, new customers get acknowledged automatically, client questions get answered by systems, staff issues get routed and summarized, decisions get flagged instead of interrupting. Marco checks one summary while on the other hand, Elena checks one dashboard.

During the hour, they decide. They accept or decline, they adjust strategy, they use leverage that actually moves the business forward. They do not hunt for low-return tasks, because the system already handled it.

During and after the hour, systemic execution continues. The follow-ups go, their staff carries out instructions, and their clients stay supported. Nobody has to wonder if something got missed.

Why This Holds During Messy Weeks

This is where it earns its keep.

A SPA therapist calls out at Elena's busiest location. Phones don't stop ringing. A client leaves a one-star review after a scheduling issue. In the old setup, Elena would spend the entire day reacting.

In the new setup, coverage rules trigger the next steps. Reviews get a calm public response and get routed for private resolution. Front desk questions get answered without escalation. Elena still decides what matters. She just doesn't absorb the chaos.

Marco saw the same effect during launches. If he skipped his one protected hour, the business slowed down, but it didn't collapse.

That's the test.

Why Focus Shrinks the Workday

Research and reporting on focused work, including Cal Newport's Deep Work concept, points to a consistent outcome: protected focus produces more useful output than long days chopped into fragments.

What I take from that is that when interruptions disappear, your work takes less time. Decisions come faster. Creative thinking stops getting squeezed into leftovers. You finish earlier with energy still in the tank. That's when life comes back online. Family dinners you actually make. Workouts you stop skipping. Afternoons off without guilt. A business that supports the reason you started it.

What Clients Notice When You Step Back

Owners worry about this part. If I'm less available, won't clients feel it?

Clients notice speed, clarity, and consistency. They notice fewer dropped balls. They notice smooth follow-up. They notice when the experience feels stable.

Delays feel careless and inconsistency creates doubt. A clear system feels like care.

The Signal That Proves It's Working

The clearest sign is simple.

You stop asking, "What should I do next?" because the day already knows. Real decisions wait for your focus hour, the noise stays in its lane, and your energy doesn't get drained by a hundred little choices.

If you're still scanning your brain for forgotten tasks, you still have open loops. That means a system is missing, not that you need more discipline.

Operating Rules That Keep It From Breaking

Systems fail when owners drift back into old habits, so we used three rules to keep the line.

When something feels heavy, ask what it's asking you to remember. If memory is required, you need a rule or a routing path.

When something gets lighter, don't refill that space but instead protect it. That's where thinking happens.

When the same decision shows up twice in a week, write it down as a rule. Rules feed systems. Systems protect attention.

What Growth Looks Like Now

Growth stops meaning longer days.

When Marco adds a new offer, he asks what decisions it introduces. When Elena adds a location, she asks what may break first.

They don't move faster than their structure. Intake rules get written, onboarding paths get clarified, support boundaries get set.

Then they scale.

What a Bad Week Looks Like Now

We will all have those weeks where our kid gets sick, you get a spike in customer complaints or order volume, you have a vacation, a partner or employee suddenly quits.

With the AI One Hour Workday OS in place as your foundation, the business slows without panic, systems keep running, nothing critical drops. The protected hour waits if it needs to.

That's resilience.

The Real Finish Line

This chapter isn't about squeezing more work into your life but it is about stopping the quiet waste. Wasted attention, wasted judgment and wasted energy.

The one hour workday is a boundary around what deserves you. When you hold that boundary, the business supports you instead of consuming you.

Next chapter, we measure what changed. The hours you reclaimed. The money you stopped leaking. The work you no longer carry.

CHAPTER 11

HOW TO MEASURE WHAT ACTUALLY CHANGED

If someone asked you today, "Did AI save you time?" Could you answer with numbers?

Marco couldn't at first and Elena couldn't either.

They both felt calmer days, fewer interruptions and better focus. That relief mattered, but it wasn't enough. Relief feels good but proof keeps systems alive.

This chapter shows how we measured what actually changed after the AI One Hour Workday OS ran in their business for a few weeks, and why those numbers were the reason the One Hour Workday held instead of slowly unraveling.

The Three Metrics That Mattered

We ignored activity metrics completely, we ignored task counts and we ignored vanity dashboards.

Instead, we tracked outcomes that changed how the day actually felt. Time returned. Dollars protected or created. Judgment preserved.

Research summarized by Harvard Business Review shows that automation delivers real return only when leaders track time saved and decision quality. My view is practical. If your calendar still fragments your attention, the system failed, no matter how advanced the tools look.

Measuring Time Returned Without Overcomplicating It

We used the same calculation for both businesses.

Handoff Hours equaled time per task multiplied by frequency. No tracking apps running all day. No timers attached to everything. We timed each workflow once and told the truth.

Before systems were in place, Marco's time disappeared in small, repeatable ways. Lead responses took about six minutes and happened roughly fifteen times a week. Onboarding questions cost about three minutes each and showed up around twenty times weekly. Call preparation and follow-up took about ten minutes per call, and content creation from scratch consumed roughly four hours a week.

All of that added up to about six and a half hours each week.

After knowledge bases and agents were running, the picture changed. Marco reviewed one daily lead summary in about ten minutes. Onboarding questions only surfaced as exceptions, costing about ten minutes a week. Call summaries took about five minutes to review, and content shifted from writing to reviewing, dropping to around ninety minutes weekly.

His total time spent fell to about an hour and a half per week.

That returned roughly five hours each week, or two hundred and sixty hours per year.

Elena's Numbers in a Multi-Location Reality

Elena's time loss looked different, but the math told the same story.

Before systems, front desk escalations alone consumed roughly three hours a day. Reviewing responses each evening added another thirty minutes. Staff questions across locations took about an hour daily. No-shows and refund handling cost another forty-five minutes. That added up to roughly six hours every day.

After my team put the AI One Hour Workday workflows in place, Elena only had to review one daily performance summary for her SPA's each day and it took less than ten minutes. True escalations took around thirty minutes. Review monitoring dropped to ten minutes, and exception handling averaged about twenty minutes.

Her total dropped to roughly an hour and fifteen minutes a day. That returned nearly five hours every day, or more than twentythree hours each week.

Turning Time Into Dollars

Time matters because of what it protects.

Marco's effective hourly value was around one hundred and fifty dollars. Two hundred and sixty hours returned per year translates into roughly $39,000 in reclaimed judgment time. He used that time to improve program retention, launch a new cohort without hiring additional support, and reduce refunds by tightening client qualification rules, and his program outcomes. Revenue increased without extending his workdays.

Elena's costs showed up differently. Her losses came from errors, manual dependency, and delays. Slow responses triggered negative reviews and inconsistent service, resulting in drops in revenue.

After I put the right systems and bots in place for Elena, her business was able to provide faster responses to customer complaints and questions, which in turn significantly reduced the amount of negative Google reviews for her SPAs.

The systemic text, email, and voice reminders reduced no-shows by 87%. Standardized service paths improved rebooking. Using conservative math, twenty-three hours per week at an estimated two hundred dollars of operational value translated to over four thousand dollars protected weekly. That's more than two hundred thousand dollars per year before any growth is added.

Research from McKinsey consistently shows that the biggest business gains come from reducing errors and delays. I believe that most losses hide in friction, not payroll.

The Quiet Metric: Judgment Preserved

The most important results never showed up in a spreadsheet.

Before the AI One Hour Workday system was in place, Marco decided almost everything by default. Elena approved almost everything by necessity. Their days were filled with small decisions that felt harmless on their own but added up to constant mental drag.

After the systems were installed, those decisions stopped landing on them automatically. Intake rules handled fit. Policies handled refunds. Defaults handled routine situations. Marco and Elena still made decisions, but only when the decision actually mattered.

That showed up in simple ways. Marco stopped replaying conversations at night and Elena stopped feeling "on call" even when she wasn't working. They both noticed they had more patience, better focus, and clearer thinking when real issues came up.

Cognitive research consistently shows that decision quality drops as choices stack up over the day. My take is practical. When you protect judgment, you don't just make better business decisions. You become easier to live with, including with yourself.

That's the return most founders are chasing without realizing it.

The Measurement System That Prevented Backsliding

They didn't bother to use dashboards. They used just one AI performance document that took five minutes a week to update.

The document had three simple columns: time saved, money protected or made, and decisions turned into rules. That was enough. No overthinking or pretending to perform.

Just the kind of proof that keeps old habits from sneaking back in.

When Numbers Change Leadership

Once Marco and Elena had numbers, something shifted.

Marco stopped second-guessing whether he should jump back into tasks that no longer needed him. He trusted the system because he could see what it returned every week.

Elena stopped apologizing to staff and clients for not responding personally. She had clear standards, clear coverage, and clear outcomes. That confidence showed up in how she enforced boundaries and how her team stepped up.

Their new metrics didn't only justify their new systems, it also changed how they showed up as leaders.

Why Measurement Keeps the One Hour Workday Real

When you can say five hours returned per week, six figures protected, or half the decisions removed, your old habits will lose their grip.

Measurement keeps the AI One Hour Workday real. It turns relief into something you protect instead of something you accidentally give back.

At this point, you've seen what changed for Marco and Elena. The time they reclaimed, the decisions they removed, and how their days started to feel lighter and more intentional.

The final question is simple: how do you do this without overhauling your entire business at once?

Chapter 12 gives you the exact five-week plan we used. One focus per week, one workflow at a time.

This is where the AI One Hour Workday stops being an idea you agree with and becomes a structure you actually live inside.

CHAPTER 12

YOUR AI ONE HOUR WORKDAY IMPLEMENTATION PLAN

If you close this book and do nothing differently next week, nothing changes.

I'm sure its not a failure of motivation but a failure of believing that your time is worth more.

Marco didn't change his workday by learning faster. Elena didn't get her time back by buying better tools. They followed an order that removed pressure first and layered support second.

This chapter gives you that order.

Follow the AI One Hour Workday implementation plan. More resources to help you are available at

www.TheAIOneHourWorkday.com. You only need one focus per week and remember, no piling on.

Week One: See Where the Pressure Actually Lives

This week is about clarity and honesty.

Complete your forty-eight-hour truth audit. Track what pulls you out of focus, not what looks productive on paper. Pay attention to what triggered the task, how long it took, and whether it required judgment.

Then map your work using the HAS framework. Decisions that only you should make stay human. Repeat work that should run without you gets automated. Drafts and preparation that require your voice become supported by AI.

Pick one workflow that shows up at least five times a week. Lead response, scheduling issues, onboarding questions, or staff interruptions all qualify. Break that workflow into steps and tag each step using the HAS framework: Human, Automated, or Supported by AI.

Your success marker for this week is simple. Install a clearly documented workflow that any member of your staff can follow.

Research summarized by Harvard Business Review consistently shows automation succeeds when leaders clarify decisions before building systems. My take is straightforward. If the rules live in your head, the system will fail.

Week Two: Build the Brain Before the Machine

This week is about memory.

Create one Custom GPT or knowledge base tied to the workflow you mapped. Load it with one clear framework, one operating procedure, five real examples written by you, and the common edge cases that usually trip things up.

Marco started with intake rules. Elena started with front desk escalation and booking standards. In both cases, they weren't asking AI to improvise but instead they were teaching it how they think.

Test the system with real questions you answer every week. If the output feels off, fix the inputs. Do not add more tools yet.

Your success marker is clear. AI drafts responses or summaries you would actually send with only light edits.

Research from MIT Sloan shows systems trained on internal standards outperform generic automation. I agree that a trained system beats manual output every time.

Week Three: Hand Off One Repeating Loop

This week is about relief.

Take the same workflow and add one simple automation or agent. One trigger. One path. One summary back to you.

This might look like a new inquiry triggering a response and qualification with a daily summary, a booking issue being resolved through a rule check and logged automatically, or a review being drafted and flagged only if sensitive.

This is where Marco reclaimed hours. It's also where Elena reclaimed sleep.

Don't aim for perfection but instead, aim for consistency.

Your success marker is that the workflow runs for three days without pulling you back in.

MIT research shows that reaching eighty percent reliability on one process beats partial automation across many. My view is blunt.

Partial relief keeps you trapped.

Week Four: Stabilize the Client or Customer Path

This week is about trust.

Map what happens after someone says yes. Onboarding, confirmations, preparation instructions, and follow-ups. Automate defaults so quality stops depending on memory.

Elena used this week to standardize service flow across locations. Marco used it to stop re-answering the same student questions.

You stay present for exceptions and the system handles what's expected.

Your success marker is one complete journey that runs end to end without you touching it.

Research from Bain & Company shows retention rises when early stage experience feels consistent. My take is simple. Clarity creates calm, and calm builds loyalty.

Week Five: Measure What Changed

This week is about proof.

Track three numbers. Time returned. Dollars protected or created. Decisions removed from your head.

Use one simple document and update it weekly. Marco saw five hours a week come back. Elena saw more than twenty. Those numbers stopped second-guessing.

Your success marker is being able to answer, without hesitation, the question "What changed because of this?"

Stanford research shows visible progress increases followthrough. I see it every time. When the numbers show relief, people protect it.

What Happens After Five Weeks

At the end of five weeks, you won't be finished. That's the point.

You won't overhaul your entire business in one sweep. You'll improve it in deliberate steps. One workflow at a time. One pressure point at a time.

Each month, look for areas where work feels heavy, and then apply the one-hour workday framework. That's how the one-hour workday holds. I don't want you to just chase it but instead, protect it.

Marco protected his time by keeping decision-making where it belonged and stopping the same questions from landing on his desk. Elena protected hers by locking in consistent systems across locations so quality didn't depend on her presence.

Different businesses. Same discipline. Both running on the AI One Hour Workday Operating System.

CONCLUSION

THE WORKDAY IS NO LONGER THE PROBLEM

Three months ago, I received a message from Tasha, a business coach who had implemented the systems in this book.

She wrote, “I forgot what it felt like to finish my workday without being exhausted. Yesterday, I closed my laptop at two in the afternoon, picked up my son from school, and we spent the rest of the day at the park. My business ran without me. I almost cried in the car because I realized this is what freedom actually feels like.” That moment speaks for itself.

Tasha didn’t wait for "the perfect time" before she redesigned the engines that were running her business. Instead, she stopped carrying her business manually and started designing it to run with AI intelligence. This shift is available to you too.

Your Business as a System

This book was written around one simple truth. You were never meant to do everything. You were meant to design how it gets done.

The AI One Hour Workday is not about working less out of laziness. I love the fact that it's about working with intention, protecting your time, preserving your judgment, and building a business that doesn't require constant supervision.

In fact, AI doesn't replace you, it removes the work that never required you in the first place. All the repetitive tasks, the manual follow-ups, the mental tracking, and the constant switching.

When that weight lifts, your attention returns to where it belongs and that is strategy, creativity, connection, and meaning.

Over these chapters, you learned how to move from operator to architect. From reacting to designing. From carrying everything to building systems that carry the work. Now it's time to apply it.

What Comes Next

The real work begins after the last page.

This isn't a finish line but instead, it is a starting point.

You already have what you need. A clear framework, real examples, proven systems, a practical implementation plan.

What matters now is action.

Just start with one workflow and choose one task that drains you every week. Automate one low-risk process, document one outcome, and measure one hour returned.

Then repeat.

Confidence grows through momentum. Time opens through consistency and that's when clarity and relief will come in.

You're no longer building through force and chaos, but you're building through design.

Your AI-Powered One Hour Workday Starts Here

It begins the moment you stop surviving your business and start shaping it.

When you protect your attention, when you let systems carry the repeat work and when you stop proving your value through exhaustion.

This isn't the end of your journey but its the beginning of a new rhythm.

Calmer mornings, manageable workdays, evenings that belong to you and your family again.

A business that grows without pulling you apart. One that respects your time and supports your life.

A Final Question

You've been given the blueprint. Now the decision is yours.

Are you ready to stop surviving your business and start designing it? Are you ready to reclaim your time and protect your energy? Are you ready to build something that works without constant effort?

If the answer is yes, start now.

Return to the beginning and identify your patterns. Choose your first handoff and automate one small task today.

Your AI one-hour workday isn't a fantasy. It's a framework.

And it starts the moment you choose to use it. Your time is yours again, your business is lighter, and your life has room for you in it.

This is just the beginning.

RESOURCES

CONTINUE YOUR JOURNEY

Your AI-Powered One Hour Workday doesn't end with this book. Here's how to take the next step:

FREE RESOURCES

AI-Powered One-Hour Workday Interactive Workbook

Step-by-step guidance, optimized prompts, and templates to help you map your patterns, document your processes, and build your first AI systems.

Download: **www.AIOneHourWorkday.com**

Weekly Newsletter: AI Insights for Business Owners

Practical AI strategies, productivity tips, and real implementation stories delivered every week.

Subscribe: **www.AIOneHourWorkday.com**

PREMIUM RESOURCES

Join the Online Community

Connect with other founders building AI-powered businesses. Share wins, get support, troubleshoot challenges, and access exclusive resources. Join the: **www.AIOneHourWorkday.com**

Work With Nicole

Apply below for white glove implementation of the AI One-Hour Workday OS inside your business. Get personalized support for your unique situation in our Founders Lab.

Apply: **www.AIOneHourWorkday.com**

Book Nicole for Your Event

Available for keynotes, workshops, training programs, podcasts, and media appearances on AI, online marketing, and women's empowerment. Contact: **www.NicoleMcCain.com**

APPENDIX A

PARTNER & STUDENT SPOTLIGHT

Inside my Digital Roadmap AI Academy and AI One Hour Workday online communities, hundreds of entrepreneurs have built their businesses using the AI systems mentioned in this book.

This appendix spotlights examples from close partners, mentors, and students who are using AI with intention and steadily building toward their own One Hour Workday. These are not theories or hype loops. This is what it looks like when strategy meets traction.

Arlan Hamilton

Award winning business coach, author, and start-up investor who created a custom chatbot called "Arlan's Homegirl," allowing students to access her guidance 24/7. The chatbot delivers grounded business advice pulled directly from her frameworks. She also builds monetizable AI-powered apps and experiences, and teaches her students to do the same.

Atena Pegler

International speaker, award winning marketer, and business coach who built an AI database capable of generating an entire website, full marketing plan, webinar scripts, and email sequences at the push of a button. Her system turns big ideas into launch-ready assets in hours instead of weeks.

Brenda Rigney

Marketing and content coach for women in midlife whose breakthrough came from building her first custom GPT. By encoding her expertise into AI tools and building over 20 specialized GPTs to support her business, she erased entire to-do lists and reclaimed 80% of her time. This creative space allowed her to strategically revive dormant offers into steady revenue drivers.

Dallas Clounch

Founder of The AI Surfer, AI coach, and content creator. Through his 90/10 philosophy, he teaches business owners how to train AI as a 90% strategic automation partner while keeping the 10% human edge that builds trust, connection, and originality. His clients average 12+ hours saved weekly. The goal is simple: humans stay in control and focus on what they actually enjoy.

Sonya Denise

Founder of Dream Development, a company that helps creatives shape and execute their ideas. Sonya has used AI note-takers for years to manage meeting notes, project summaries, and scopes of work, helping her surface next steps instantly. The result: faster decisions, stronger momentum, and significantly less busywork standing between ideas and action.

Daisy Jones

Business and mindset coach who once carried most of her business in her head. AI became her creative partner, helping her clarify her voice, build her own pages and tools, and stop outsourcing everything. She now saves 10 to 15 hours per week, moves faster, shows up with confidence, and lets her expertise actually do its job.

These stories aren't outliers, they are examples of what becomes possible when you stop managing tasks manually and start designing intelligent systems.

Your story could be next.

www.TheAIOneHourWorkday.com

APPENDIX B

DEFINITIONS & KEY TERMS

Core Frameworks from This Book:

48-Hour Truth Audit

A tracking exercise (Chapter 2) where you log every task, tool switch, and interruption for two days to reveal your actual work patterns versus your perceived patterns. Shows where time really goes and identifies automation opportunities.

Handoff Number

Your calculated hourly value (annual take-home ÷ 2,000 ÷ 4) used to determine which tasks to delegate or automate. If a task can be handled for less than this number, it should be handed off. Introduced in Chapter 2.

Memory vs. Judgment Tasks

A categorization framework (Chapter 1) that separates work requiring your unique human judgment (strategy, creative thinking, high-leverage decisions) from work that simply needs to happen consistently (follow-ups, routine responses, tracking).

AI One-Hour Workday

The amount of time your business should require from you for work that depends on your unique human judgment, strategy, decisions, effort, creative thinking, and high-leverage conversations. Everything else should run through systems and automation.

The AI One Hour Workday Operating System

Nicole McCain's proprietary system of AI agents, workflows, and decision rules that replaces manual oversight. Implemented to execute daily operations, drafts responses, routes issues, and enforces standards so the owner's time is reserved for one hour of strategic, human-only work.

Three Engines Framework

The three core systems every business needs (Chapter 3): **Growth** (turning interest into income), **Fulfillment** (delivering outcomes to clients), and **Innovation** (improve-ment and adaptation). When these engines run without constant manual intervention, business becomes sustainable.

Artificial Intelligence Terms

Artificial Intelligence (AI)

Software systems designed to perform tasks that typically require human intelligence, such as writing, summarizing information, recognizing patterns, making recommendations, and supporting decision-making. In this book, AI is used as a practical support layer that carries repeat work, holds context, and reduces mental load so your judgment is preserved.

→ First mentioned in: Introduction

→ See also: Large Language Models (LLMs), Agentic AI, Custom GPT, Automation

Artificial General Intelligence (AGI)

A theoretical form of AI capable of reasoning, learning, and applying knowledge across many domains at a human-level or beyond. AGI would be able to transfer understanding from one area to another without task-specific training. This book does not rely on AGI. All examples and systems described use today's narrow, task-focused AI designed to support specific workflows, not replace human judgment.

→ First mentioned in: Chapter 4 (context-setting and clarification of AI limits)

→ See also: Artificial Intelligence (AI), Agentic AI, Human-Only Decisions

Agentic AI

AI that can take action, make decisions within boundaries, complete multi-step tasks, and self-correct. Think of it as an automated digital assistant that executes workflows autonomously.

→ See also: Automation, Workflow

Custom GPT

A trained version of an AI assistant that learns from your own data, knowledge base, style, and preferences so it can respond and output information aligned to your writing style, voice, and frameworks.

→ See also: LLMs, Tone Guide

Generative AI

AI systems that create content—text, images, or other media—based on prompts. In this book, this includes text generation and multimodal models paired with image generation capabilities.

→ See also: LLMs, Multimodal AI

LLMs (Large Language Models)

Text-based AI models trained to understand language, write, plan, outline, summarize, and brainstorm. Examples include GPT-4, Claude, and Gemini.

→ See also: Custom GPT, Generative AI

Multimodal AI

AI that can interpret multiple types of input—text, audio, images, PDFs, screenshots—and sometimes generate visuals depending on the model.

Enables more flexible and context-aware automation.

→ See also: Generative AI

Prompt

A written instruction you give to an AI model telling it what to create, consider, or analyze. Effective prompting is key to getting useful AI outputs.

Business & Automation Terms

Automation

A system that performs tasks automatically based on rules or triggers without human intervention. Reduces manual work and ensures consistency.

→ See also: Workflow, Agentic AI

Content Studio

A workflow that turns one idea into multiple content formats using AI-powered drafting, cleanup, visual generation, and scheduling. Maximizes content leverage.

→ See also: Workflow, Pipeline

CRM (Customer Relationship Management)

A tool that stores client and lead information, tracks conversations, manages pipelines, and organizes follow-ups. Essential for Growth engine automation.

Lead

A lead is a person or business that has shown interest in your product or service. This can happen when someone fills out a form, sends a message, books a call, signs up for a list, or contacts you with a question. A lead is not yet a customer, but they have raised their hand and may become one if followed up with properly.

Pipeline

A sequence of steps that a piece of content or task travels through, from idea capture to publishing or completion. Visualizing pipelines helps identify automation opportunities.

→ See also: Workflow

SOP (Standard Operating Procedures)

Documented step-by-step instructions for completing recurring tasks. SOPs are essential for training AI and team members to execute work consistently.

→ See also: Workflow, System Architecture

System Architecture

The structure, workflows, and logic that define how your business operates behind the scenes. Good architecture reduces dependency on any single person.

→ See also: Three Engines Framework

Tone Guide

A document that describes your voice, emotional style, pacing, humor, boundaries, and tone preferences—used to train AI to write like you while maintaining authenticity.

→ See also: Custom GPT

Value Engine

The core process in your business that creates growth, fulfillment, or innovation. These engines determine how your business produces revenue and results.

→ See also: Three Engines Framework

Workflow

A series of steps that turn an input into an output. Your "behind-the-scenes" process for getting something done, whether manual or automated.

→ See also: Pipeline, Automation

APPENDIX C

SOURCES & FURTHER READING

The research and frameworks in this book draw from academic studies, industry reports, and leading experts in productivity, AI, and business operations. The sources below provide deeper exploration of key concepts.

Productivity & Workplace Research

Burnout and Workload in Entrepreneurship

University of New Hampshire (2024). "Burnout among entrepreneurs and micro-business owners: Stress, workload, and work–life balance challenges."

→ Referenced in: Chapter 1 (Handoff Section)
Available:https://scholars.unh.edu/cgi/viewcontent.cgi?article=1141&context=ms_leadership

Work Interruptions and Task Complexity

National Library of Medicine (2024). "Work interruptions of office workers: The influence of the complexity of primary work tasks on the perception of interruptions." PMC10789356.

→ Referenced in: Chapter 2 (context switching costs)

Available:https://pmc.ncbi.nlm.nih.gov/articles/PMC10789356/

Small Business Time Management Study

Salesforce and Slack (2024). "Small Business Owners Lose 1.5 Hours Daily to Wasted Time." Slack Survey Findings.

→ Referenced in: Introduction (96-minute productivity loss)

Task Resumption After Interruption

Mark, G., Gonzalez, V.M., & Harris, J. (2005). "No Task Left Behind? Examining the Nature of Fragmented Work." Proceedings of CHI 2005, 321-330.

→ Referenced in: Chapter 1 (23-minute recovery time after interruption)

Lead Response Time and Conversion

Oldroyd, J.B., McElheran, K., & Elkington, D. (2011). "The Short Life of Online Sales Leads." Harvard Business Review.

→ Referenced in: Chapter 3 (response time impact on lead conversion)

Technology Adoption and Organizational Productivity

Brynjolfsson, E., & Hitt, L.M. (2003). "Computing Productivity: Firm-Level Evidence." Review of Economics and Statistics, 85(4), 793-808.

→ Referenced in: Chapter 2 (technology adoption patterns and productivity gains)

Artificial Intelligence Foundations

History of Artificial Intelligence

Stanford University (2023). "Stanford Artificial Intelligence Laboratory: History and Milestones."

→ Referenced in: Chapter 4 (AI evolution timeline)

Available: https://ai.stanford.edu/

Deep Learning Breakthrough

Krizhevsky, A., Sutskever, I., & Hinton, G.E. (2012). "ImageNet Classification with Deep Convolutional Neural Networks." Advances in Neural Information Processing Systems, 25, 1097-1105.

→ Referenced in: Chapter 4 (neural networks and pattern recognition)

Large Language Models

Brown, T., Mann, B., Ryder, N., et al. (2020). "Language Models are Few-Shot Learners." OpenAI GPT-3 Technical Report. arXiv:2005.14165.

→ Referenced in: Chapter 5 (how AI learns patterns and context)

Available: https://arxiv.org/abs/2005.14165

Business Applications of AI

AI and Productivity Gains

MIT Sloan Management Review (2023). "Generative AI and the Future of Work: How AI Will Transform Business Operations."

→ Referenced in: Chapter 6 (AI productivity multipliers in knowledge work)

Economic Impact of Generative AI

McKinsey Global Institute (2023). "The Economic Potential of Generative AI:

The Next Productivity Frontier."

→ Referenced in: Chapter 8 (business automation ROI and economic impact) Available:https://www.mckinsey.com/capabilities/mckinsey-digital/ourinsights/the-economic-potential-of-generative-ai-the-next-productivity-frontier **AI in Customer Service**

Harvard Business Review (2023). "The New Frontier of AI-Powered Customer Experience."

→ Referenced in: Chapter 8 (automated customer support and service quality)

Customer Experience and Retention

Bain & Company. "The Value of Online Customer Loyalty and How You Can Capture It."

→ Referenced in: Chapter 12 (consistency in customer experience and retention)

Future of Work Research

Microsoft (2024). "Work Trend Index 2024: AI at Work Is Here. Now Comes the Hard Part."

→ Referenced in: Chapter 11 (measuring AI adoption and workplace transformation)

Available: https://www.microsoft.com/en-us/worklab/work-trend-index

Visible Progress and Motivation

Amabile, T., & Kramer, S. (2011). "The Power of Small Wins." Harvard Business Review, 89(5), 70-80.

→ Referenced in: Chapter 12 (tracking progress increases implementation success)

ABOUT THE AUTHOR

Nicole McCain is an AI business strategist, speaker, trainer, and consultant who helps high achieving women and small business owners scale with AI while reclaiming their time and building businesses that deliver both profit and freedom.

With over twenty years leading multi-million dollar initiatives inside top S&P 500 companies, Nicole brings enterprise grade strategy and lean productivity discipline into the world of modern entrepreneurship. She is the founder of The AI One Hour Workday Collective, The Digital Roadmap AI Academy, and Gentiv AI Consulting, where she designs AI driven systems that cut busywork, protect focus, and create measurable revenue lift.

Nicole helps founders stop managing chaos and start running businesses that operate with intention. Her work has supported hundreds of founders to remove repetitive tasks, stabilizing their marketing engines, and building companies that continue to perform without constant oversight.

The outcome is simple and provable. Fewer hours lost to noise. Stronger execution. More control over time, energy, and growth.

When she is not coaching, designing client workflows, or vibe coding new AI APPs, Nicole enjoys spending time with her family at the beach and serving in her church.

Monthly Support

Our membership gives you live training, coaching, and a community of successful business owners mastering and scaling with AI.

www.TheAIOneHourWorkday.com

The AI One-Hour Workday Founder's Lab™ Waitlist

A 4-week small group intensive where we help you to implement AI into your business. We build this with you.

www.TheAIOneHourWorkday.com

www.ingramcontent.com/pod-product-compliance
Lightning Source LLC
LaVergne TN
LVHW050824061126
851857LV00012B/972